CW01430091

For
my Grandparents

John & Jean
Neil & Janet

Jenna Whyte
2 Cambridge Drive
Leeds

The Illustrated Guide to the Elements
Volume 2

First published in
Great Britain 2014
By
Jenna White

All rights reserved.
No part of this publication may be reproduced, stored as electrical data, hired out, or transmitted in any form or by any other means without
prior permission in writing of Jenna White.

Any enquires concerning reproduction of this book should be emailed to
Jennawhyte@hotmail.com

ISBN
978-0-9575327-1-7

www.Jennawhyte.co.uk

First Edition
15/03/14

Cover illustration by Jenna Whyte
Cover design by Holly Trafankowska

The Illustrated Guide to the Elements

Volume ii

by Jenna Whyte

COBALT'S

...§ Finest §...

Invisible Ink

One can use this ink to record the
the most wicked thoughts and

malicious statements

YOUR SECRETS ARE
SAFE!

....Until Heated

!The Most Wonderful Product Of It's Age !

For rebel meeting locations, love.
trysts, plots of an unsavoury nature
or a good old fashioned X to mark the
spot.

Astonishes and delights all who use

Elements Volume II
Contents

Dear Readers,

In Volume ii the investigation into the electron infatuated world of the Elements continues. We concentrate this edition on the Alkaline Earth Metals and the largest of all groups; the Transition Metals.

As with Volume i we will take an all-pervading exploration of the above mentioned groups with the same degree of detail and professional integrity. It is my endeavour to discover the methods they employ to exchange their electrons and protect themselves from the perils of industry and understand how their physical properties influence their personalities and relationships, in short to uncover the peculiarities of each Element. Once again each profile will include a hand painted likeness of the Element in question.

As we scrutinize these families we see that the Alkaline Earth Metals bear remarkable similarities to the Alkaline Metals (as researched in Volume i) and are some of the most distinguished and influential Elements in our world. While we owe our lives to many, others are accountable for countless deaths - intentional or otherwise.

We will also observe the most extensive of the Element groups; the Transition Metals. They form almost 75% of the entire Elemental world and infiltrate our lives; physically, physiologically, biologically and in some cases even spiritually. On looking further into this vast family I have uncovered sub-fractions or group divisions. Thus I have decided to add sub-chapters to better aid understanding of Elemental society by breaking down this substantial chapter. These sub-chapters focus on the Coinage Metals the Platinum Group and the Refractory Metals.

Volume ii has revealed as many surprises and revelations as the first compendium, these being equally engaging. It is my aim that by examining the facts you will see just how much these remarkable women influence our lives, dictating many of the rules by which we must abide.

Jenna Whyte

Jenna Whyte

THE
ILLUSTRATED GUIDE
TO THE ELEMENTS
VOLUME I

FIND ALUMINIUM AND HER COLLEAGUES IN THE ENGAGING INVESTIGATION INTO THE PERIODIC TABLE: TAKE A THRILLING AND INSIGHTFUL LOOK IN TO THE HALOGENS, ALKALINE METALS, METALLOIDS, POST TRANSITION METALS, NOBLE GASES AND THE NONE-METALS!

!With Hand Painted Illustrations!

The Alkaline Earth Metals

The Alkaline Earth Metals can be somewhat of a Jekyll and Hide kinfolk. As we look into their activities we discover how hard they work to better our human lives and control their impetuses. The next moment however they exhibit animal like behaviour, fits and most unbecoming violent traits. Like all the women in their Elemental world, the primary objective is to have a full outer shell of electrons. In this case they have two electrons in their outer shell and the quickest and easiest way to achieve a complete shell is to rid themselves of the two offending electrons. As a result they can suffer from angry outbursts in a bid to eradicate the unwanted cargo. In the main this causes them to be sensitive and dangerous to the naïve or those not armed with the appropriate knowledge or protective gear.

However it is thanks to their reactive natures that they have become an integral part of the humanoid world. As these Elements find it hard to cope with their 'condition' they heavily rely on other Elements to take their electron load. They make alliances, coalitions and deals with other Elements who are in need of extra electrons, and create valuable working partnerships. We will uncover the extent of how far these partnerships' effect our lives in individual case files.

The Alkaline Earth Metals are close cousins of the Alkaline Metals and share many of their characteristics (See Volume I for more on the Alkaline Metals), yet these conscientious Elements have not enjoyed the same privileged background. Thusly they have had to become tougher; The Alkaline Earth Metals are less reactive than their volatile relatives, not as tainted by the outside world, and in many cases have found ways to curb their tempers and idiosyncrasies, turning them into productive industries or ways to safeguard themselves. Be that as it may, they are still softer than other Elements and unpredictable at times.

Much like their close relatives, The Alkaline Metals, the Elements further down the column presented in the periodic table become more belligerent as the electrons face less pull from their nuclei (see extra notes), due to their bigger mass and greater electron shielding. This means the outer electrons find it easier to leave and the Element feels the need to rid themselves even more of the errant electrons they can't control. The less they can control their electrons the less they can control themselves.

There are some substances or Elements which will undoubtedly unleash the group's volatility and hysteria. For example all in this family will react with water with varying degrees of madness, and the family of Halogen (also analysed in Volume i) will inevitably bring out the worst in them. Perhaps the most significant catalyst to their dark psyche is fire, to which they all react explosively and with bright flashes.

But all in this brood are fighting to keep their condition in check. In many ways this distinguishes from the Alkaline Metals who have little or no control over their actions.

Beryllium 4

Name: from the Greek 'beryllos' meaning the mineral beryl or "earth of beryl".

Subject Notes:

Beryllium likes to give the impression she is innocent, sweet and vulnerable. Nothing could be further from the truth regarding this Element. Behind her façade of shiny gemstones and honeyed tongue she is a highly toxic lady. If one were foolish enough to be lured in, they may find themselves suffering from "Berylliosis" (see 'subject history' for details) or worse, simply deceased.

She is more than happy to pay a few electrons to keep her image intact, but she is as likely to take twice what she is willing to give.

This wolf in sheep's clothing has also grown apt at looking after her self. She forms a protective oxide layer that can shield her from air, water, nitric acid and even red heat, fighting off Oxidization at temperatures up to 600 ℃.

Subject History:

Miss Beryllium is never found alone but usually surrounded by a host of precious gemstones like aquamarine, emeralds and chrysoberyl. This is all part of her ploy to entice and deceive. If this is not enough she is also sweet to taste. When she was first found her discovers wanted to name her "glycenum" or "glucina" from the Greek meaning sweet.

Association with this Element is corrosive to the sprit as well as body tissue. If inhaled a life threatening allergic reaction will take place as named above -'Berylliosis'. Symptoms include inflammation of the lungs, shortness of breath and a nasty cough. It can take a long time to show its symptoms and a third of those infected by Element 4's sickly sweet infusion will die. The 'lucky' survivors will be permanently disabled. If ingested she is slightly less toxic but can still kill and cause plenteous damage to the human body. She finds this toxic jaunt on the body both diverting and relatively easy. The reason for this ease is down to her relationship with Element 12: Magnesium, because they are in the same family they have similar traits and tendencies. However whereas Magnesium is vital to human life and keeps the body running Beryllium does not. The body mistakes her for Magnesium letting her in where she pleases. Once inside she displaces Magnesium and starves the body of her vital help.

Luckily for all Beryllium is a very rare Element not just on earth but also in space.

Magnesium 12

Name: Named after 'Magnesia'- a district in eastern Thessaly, Greece (which was named after the 'Magnetes' tribe)

Subject Notes:

In biological and industrial roles Magnesium is well sought after. Her ever enterprising work ethic places her at the very centre of all life on earth and marks her as one of the 'Titans' of industry in the Elemental world.

The Alkaline Metals endure what some may view as 'disabilities'. Their soft natures, prone to violent and explosive outbursts, are easily affected by their surrounding conditions. There are those, however, who can turn these 'disabilities' into something beneficial; Magnesium is one such Element. Not only does she have this skill but has developed ways to protect herself from her own frailties and volatile nature.

In a bid to rid herself of two electrons Element 12 is capable of fierce fits, not only combusting but also letting off the brightest of white lights capable of blinding those foolish enough to gaze at the spectacle of her ignition. Magnesium lives in fear of such episodes and endeavours to control her temper in a solid mass form. In this form she becomes almost impossible to ignite. Only when in small forms will she submit to the heat and dissolve into hysterics. Once in this state of aberration she cannot be subdued until the fit has run its course. She will continue to burn without the presence of Oxygen, and if doused in water she will burn more ardently. Covering her in sand has proved to be the most effective method of neutralization.

This aside Subject 12 proves to be tough, excelling in the art of self-preservation. In defiance of her 'genetics' Magnesium has found ways to be stable in air without the aid of oils or incarceration. She does this by using the powers of Oxygen, creating an Oxide layer which protects her from the outside world. As the result of such genius Magnesium and her alloys become corrosion resistant.

Subject History:

Magnesium is a huge player in the industrial world thanks to the abilities listed in above, namely her light weight and anti-corrosion properties. Her charges include photography, flares, aircrafts, automobile engines, missile construction and incendiary bombs which allowed wide spread firestorms that if left unattended may engulf whole cities. However she has biological roles that affect all living matter and are far more significant to us as humans. She regulates movements through membranes, helps maintain bone structure, builds proteins, is involved in replicating DNA and is a co-worker in over 100 different enzymes.

This multi-talented Element is also at the very heart of the chlorophyll molecule and thus responsible for the air we breathe and food we eat. She causes the chlorophyll to absorb the blue and red wavelengths of light but not the green.

Magnesium, like her Alkaline relatives, has a dim and gloomy outlook on life and is constantly battling her own bouts of depression, genetic disorders and assaults from other Elements and conditions. Despite this she does rather well at keeping herself in check and ever professional, does not let her mood swings affect her work.

Calcium 20

Name: from the Latin 'Calx' meaning lime.

Subject Notes:

Calcium is something of a mad genius. Suffering from the same electron condition as the rest of the family she is highly reactive and restless. This results in questionable sanity and eccentricities. However, unlike many Elements I have investigated, Calcium uses her reactive temperament to draw others to her and makes productive alliances that work to enhance her enterprises, of which there are many (in fact Element 20 is the 3rd most abundant Metal on the Earth). She bonds with ease to other Elements, with her gentle but unremitting nature and is well liked in the Elemental world. Her brand of madness is often over looked because she is very useful, but more importantly she rewards her associates generously with her extra electrons. Subject 20 is hungry to be the centre of attention, showing her brilliance on every available occasion, often causing spectacles and resplendent scenes with her trademark brilliant white products for which she has worldwide acclaim.

However behind her showy white smile and theatrical performances, she is, deep down, like the rest of her family: soft natured and easily hurt whether it is via general rough usage or attacks from acids or water.

Subject History:

As mentioned in her Subject Notes, Calcium's domain is vast and is commonly recognised by her white signature. Automatically many think of bones, teeth, milk, chalk, marble, shells and sometimes coral reefs. This, is just the beginning of her works, for example the white cliffs of Dover are attributed to Subject 20. Furthermore the Romans used lime (CaO) to make cement for building amphitheaters and aqueducts (it is also known as 'quicklime', which in turn can be converted to 'slaked lime' by adding water which was used to neutralize acid in soil). The list continues, historically she has been flung into graves and sewage, and used in the treatment of drinking water (especially for hardening and Arsenic removal from water).

Perhaps most importantly she is vital to almost all life. Calcium ions are important messengers between cells in humans, plants and animals and are absolutely essential for the existence of multicellular life forms. Not only that, she is the architect of our bone structure, she keeps the pH of our blood stable, helps the clotting of our blood and triggers hormone release.

It has been stated that Element 20 enjoys the spotlight where possible. She adores it so much in fact that she coined a well-known phrase - when lime burns in an Oxy-Hydrogen flame she lets off an incandescent glow. This was used to light the stage in theaters during the 1800's until electricity took over – hence the saying 'Limelight'.

Ca 20

Calcium

Strontium 38

Name: Named after the town Strotian in Scotland.

Subject Notes:

Miss Strontium, known for her fiery highland temperament, has had an extended stay at the 'Asylum for Electron Challenged Elements'. This incarceration is due to her perpetual rage, fiery temper and her flair for pyrotechnic artistry – or arson. She is easily vexed by all around her; a naked flame, while she is out in the open, can easily ignite her already excitable nature, or the smallest body of water will incite fits and tantrums of epic proportions. For this reason the Asylum provides specially prepared lodgings for Element 38. By wrapping her in paraffin they hope to protect her from anything that may engender such heated rapscallion like behaviour. Judging by her decorum and records, her stay at the Asylum will be long if not indefinite, safeguarding the public from her violent attacks and cantankerous moods.

Subject History:

The human body does not require her assistance, but will often tolerate her presence, mistaking her for other more helpful Elements of a similar biology (like Calcium or Bromine). In fact a large amount of Strontium is deposited in our bones; luckily she is not of a toxic nature.

Very rarely however, when humans tamper with her, she encounters a mood so vile she does become harmful. Isotope Strontium -90 is a B-emitter that can cause serious damage to dividing cells. With a half-life of 29 years children, if exposed, will suffer her damaging presence for almost all of their lives. As stated this is rare, and is restricted to some nuclear reactors and power plants, thus not a generally a danger to the public.

It is due to Patient 38's depraved and unpredictable nature we have few uses for her. Her main occupation is in fireworks, where her anger can be channelled. Strontium is renowned for her vibrant and bold red flame in fireworks and no other Element could ever replace her stunning vermillion red.

Barium 56

Name: From the Greek 'barys' meaning 'heavy', because her minerals are as described.

Subject Notes:

Patient 56 is a permanent resident in 'The Asylum for Electron Challenged Elements' due to her volatile, malevolent and morose nature. This uncivilized creature has a plethora of unsavoury habits which make it necessary to incarcerate her for the good of the general public and, more importantly, her own safety.

If exposed to flame Barium will react violently unleashing a torrent of hatred on the world, exposing her inner most feelings in a most disturbing manner. All Subject 56's compounds that are soluble in water or acid prove to be extremely poisonous. These salts are so deadly because they are easily dissolved in the stomach acid, giving her free access to the body.

She also reacts hysterically when faced with water or alcohol (like Lithium) and does not react well to the outside world (much like her Alkaline sisters). The usually silvery Metal oxidizes expeditiously becoming black of heart and soul. Subject 56 should be stored under petroleum or other Oxygen-free liquids to prevent further depression of this austere Element.

This deeply troubled Element also shows other signs of an unstable mind, in that she has a great fear of the dark. So much is this foreboding prevalent in her life she has developed the ability to glow in the dark. If left in the sun or heated the 'Bologna Stone' – as she has become known in many circles- will glow into the night, thus relieving her of any 'dark' induced stress that may cause an attack of some kind.

Subject History:

Barium is best known when supervised by Sulphur in the so called 'Barium Meal'; this is a procedure in which radiographs of the oesophagus and stomach are taken after Barium Sulphate is ingested by a patient. The gastrointestinal tract, is a soft-tissue structure and as a consequence does not show clearly enough for diagnostic purposes on plain radiographs, but Barium's salts are radiopaque (meaning they show clearly on a radiograph) and if taken before radiographs, Barium, within the oesophagus or desired area shows the shape of the lumina of these organs. This makes a diagnosis much easier to achieve.

Subject 56 is not vital to human life and indeed because of her unreliable complexion and unsavoury attitude has very few uses. Any effects she may have on the body are of an undesirable nature. For example: Barium is much like Calcium in her biology so is easily admitted into the body. Too much may stimulate the metabolism causing ventricular fibrillation or an irregular heartbeat. As stated above her soluble salts are extremely toxic and may cause vomiting, diarrhoea, colic, tremors and paralysis.

This leads on to one of the few jobs for which she is well suited given her heinous temperament, that of rat poison.

Her other works include providing a vivid green in fireworks and scavenging air in vacuum tubes. There are few other services that Barium provides.

Real Artists Would DIE FOR THEIR ART

THE Cadmium Paint Company

May contain: Cadmium, Lead, Arsenic and other toxic Elements

price per tube 2'9   POST FREE

WARNING: Could Cause: HYSTERIA, IRRITABILITY OF TEMPER AND A PAINFULL DEATH!!!

The Transition Metals

The Transition Metals make up over 75% of all Elements and like any other group they have properties and characteristics in common. These qualities stem from their genetic makeup or atomic structure (see fig.2).

All Transition Metals have free electrons, often called the 'sea of electrons'. Because of these free electrons they can carry currents, enabling them to be good conductors of electricity and heat, furthermore they allow the outer electron of another atom to move freely, giving rise to their malleability, as each atom can slide over each other with ease. In addition because they all have these free electrons in their outer shell they are at liberty to make alloys or alliances with whom they choose - proving useful in their industrial world.

This is not the end to their genetic gifts; as their atoms are held tightly together in a military, regular structure it empowers them to be strong and dense.

These points help explain what physiognomies bind them together as a group.

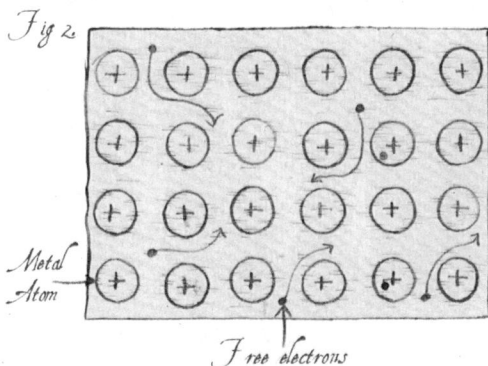

Fig 2.

Metal Atom

Free electrons

The Transition Metals are an extremely hard working party. Their assets make them perfect for industry and they toil together in their working class state to enhance each other's abilities, strengthen their weaknesses and generally work for a brighter tomorrow. As we will see however, much like our own Victorian era, whilest most are of working class and will accept almost any job no matter how insignificant or seemingly demeaning it is, there will always be those groups who are richer and think themselves above the average Element demanding more respect and authority. Whenever there is an assemblage as large as the Transition Metal group there will always be power struggles, cliques and syndicates fighting for prominence and prestige. I have in my inquiry found several, but have been able to separate them into three main groups. They each will have their own sub-chapter to help better understand their mind-sets and political views and give them their due attention. They are as follows 'The Coinage Group', 'The Refractory Metals' and 'The Platinum Metals'.

Scandium 21

Name: Derives from 'scandia' the Latin name for Scandinavia.

Subject Notes:

Subject 21 is not a celebrated Element, nor is she exceedingly rare, and few may have heard of her. She is sometimes classed as a 'Rare Earth Metal', though this can be a deceptive description given that she is in fact more abundant than Tin on Earth; the classification refers to the fact that she distributes herself sparsely in trace amounts and has her fingers in many different minerals making her hard to locate. This has given her the reputation in the Elemental world as an urchin, appearing in places where she is unwelcome and never making herself abundant enough to be of use.

An interesting fact regarding Element 21's abundance is that she is more plentiful in the heavens than on Earth. It may be that she considers this to be one of her bolt-holes where she can avoid too much labour and stay on the outside ring of society; though this is only conjecture.

While Scandium is by no means a weak Element in comparison with other Transition Metals she is considered to have more of a skittish nature and has what many may perceive to be weaknesses. For instance she is susceptible to weathering, will ignite in air, react with water and dissolve in almost all acids.

Subject History:

Though Demetri Mendeleev predicted her existence, Scandium was only discovered in 1879 and it took another 58 years for us to learn how to manage her, as she is exceptionally difficult to 'prepare' for any kind of work. Indeed any practical application came 33 years later in very specialized fields like neutron filters in nuclear reactors and seed germination.

Though a reprobate she does hold potential which is yet to be unlocked. She is much like Aluminium and in many ways superior, if she would only exert the same effort. Scandium is as light weight as Aluminium but has a melting point 900 °C higher. But in the absence of a stable and reliable source or a willing, agreeable nature, her production is limited to a mere 50kg a year worldwide.

At one time the great English pastime of cricket was influenced by her rapscallion ways when producers of cricket bats started to include Scandium, claiming that she lent the bats better striking power. This was eventually considered too un-sportsman like and the 'improvements' were then banned from the games. There is no evidence to support the notion that Element 21 makes a better hitting bat but this serves as an obliging reminder of the antics Scandium likes to get up to and how mischievous she can be.

Titanium 22

Name: Named after the Titans, sons of the Earth Goddess in Greek mythology.

Subject Notes:

Titanium has all the strength of steel but is 45% lighter, the highest strength-to-weight ratio of any Metal, and has shown extraordinary resilience in the face of extreme conditions. Further inspection gives a familiar yet remarkable explanation for such aptitudes in the form of an Oxide layer. This layer is initially thin but can grow over time becoming thicker. This however is not that unusual, more intriguing is the way Titanium can control the Oxide layer; if Subject 22 becomes wounded or damaged the layer can repair itself and protect the Element from further damage. She also has considerable advantages as a Metal as she does not suffer from Metal fatigue. She will graciously work with other Metals and stick to them in times of need. Titanium has her share of liabilities however and is not the super Metal she would like all to believe. She suffers from low conductivity of both heat and electricity. She might be lighter than steel but she lacks the elasticity and she shows weakness in the presence of Hydrogen. Nonetheless she is a valued Element to both the Elemental and humanoid worlds.

But what of her personality? Although often cocky and with a confidence bordering on arrogance, we can see from her working alliances that she is an Element that gets on easily with others, has a calm head, but will not subdue to the peculiarities of other Elements. Her involvement in jewellery and paint also suggests a streak of showmanship, a tendency toward grandiloquence and no small amount of swagger.

Subject History:

Element 22 is a vital asset to the Elemental military. Her foremost occupation sees her in the transport side of operations as a Captain. This roll is mirrored in our realm; because of her relative chemical inertness she is cool under pressure and her light-weight alloys make her perfect for aeroplanes and ships of all kinds. Her responsibilities include fan blades on engines, hulls of aircraft, propeller shafts, hulls on submarines, rigging and countless other parts of ships exposed to sea water which require corrosion resistant properties. Titanium has an in depth knowledge of aeronautical design and ship engineering.

Impressed with her work, other industries have put her expertise to good use. Seeing the need for higher quality pipe systems in power plants (that must be corrosive resistant) she has taken over this enterprise. But Element 22 also has an inventive side as well as being a tough military leader of Elements. Titanium's durability has become popular in designer jewellery. Her inertness and composure has once again been shown worthy of responsibility making her a good choice for those who suffer with allergies or those who will be wearing the jewellery in tough environments. Titanium dioxide has also been the saviour of artistes everywhere, replacing Lead white paint. Unlike Lead white paint which blackens over time ruining the painting Titanium's paint remains a bright white with a high reflective index and does not have a toxic nature; which is always a plus.

22

T

Titanium

Vanadium 23

Name: Named after Vanadis – another name for Freyja – the Scandinavian Goddess of beauty, or 'Lady of the Beautiful People'.

Subject Notes:

The link between Element 23 and her name is no flight of fancy or happy accident, thus it is worth looking deeper into the myth of Vanadis to help get a full picture of the Element in question. It was said that Vanadis was a goddess of love, beauty, fertility, war and death, and when she wept her tears were a striking red and gold (in fact it was believed they were the source of amber). This fable is fitting on many levels as we will discover. For a start Subject 23's salts when heated become the most vibrant, deep red worthy of Vanadis' tears explaining, consequently, how she got her name.

The likeness of this Element with the goddess does not end with red salts. While Subject 23 is an undoubtedly stunning Element she works within the realm of the Element Military as her main vocation. She brings a touch of artistry, beauty and refinements to the army with her abilities to be strong, handle high temperatures and not 'creep'. She is matchless at enhancing machinery, weapons and armour with the lightest of touches. She does not feel the need to dress in the dowdy and uncomfortable uniforms of the Element militia as some do, but prefers a garment of a more elegant and pleasing nature. This pretence however should not be confused with a soft heart or a vice of the vain as we will determine in 'Element History'.

She has antipathy for rainy unpleasant weather, when not alloyed with other Metals which causes her some distress. This is due to the fact she is very soluble, as a result the weather is the key to the distribution of Vanadium in our environment.

Subject History:

Like the goddess of war and death Vanadium has contributed strongly to the military and its enterprises, her insights and innovation are invaluable whether it be in defence or in weaponry. Her presence helps steel to be stronger and lighter, enabling the fitting of cannons on aeroplanes. Other weapon based accomplishments owed to Vanadium include the beautiful and resilient 'Damascus Steel', famous for not only its strength and durability, but also its light weight and capacity to hold a lasting sharp edge. Additionally Vanadium feels a strong sense of responsibility to her comrades and those in her charge. She thus replaces Nickel in items like helmets and war machines, glad to have the chance to protect them.

Just as there are two sides to her namesake the same is true of this Element and her accomplishments; the strong and fearless warrior produces fine arms and brings her ingenuity to the battle field and the more artistic being enjoys beauty and endeavours to bring splendour into all she does. Indications of this artistry lay in her involvements in ceramics on which her multi-coloured salts produce a regal gold colour, and blues and greens in ornate glass work. Her foremost aim appears to bring beauty and colour into a drab world constantly in conflict.

23

V

Vanadium

Chromium 24

Name: Derived from the Greek 'Chroma' meaning colour. Salts of Chromium are often striking colours.

Subject Notes:

Element 24 is glamorous, polished, stylish and a performer. When she was discovered around about the same time as Aluminium in 1761, she shot to stardom with her sleek shine. Being impervious to Oxidization, her exterior became 'The Plating' on everything from cars to cutlery, works of art and more. Much like the actress Sarah Bernhardt (1844 –1923) from our world, Chromium knows that appearance is everything and one must always be in the public eye to be in their hearts. She has thus made a name for herself not only for glamour, but also for being a modern, speedy, exciting Element, the Element of the future.

She does, however, also have another reputation, that of being a diva. She not only revels in her glossy shine but also has a fierce love for vibrant colours and gemstones and would never be seen unadorned without said stones (see Subject History). Another aspect of her prima donna behaviour is her toxic tendencies. Those unfortunate enough to spend too much time with her may develop 'Chrome Ulcers' which as the name suggests, involves ulcers or 'holes' in the skin which expose raw flesh resulting in an unbearable itch.

Though she would wish to blind her public with all her glitter, flair and talent, she, like many in the world of performance could also be regarded as superficial and shallow. Unlike Aluminium her fame was mostly achieved by plating other metals, suggesting that behind the allure is nothing more than cheap plastic or other inferior materials.

Subject History:

The history of Subject 24 has been a rollercoaster. Her fame has flitted from one scene to the next, loved and loathed by the fickle public. By 1827 she was the height of fashion and a symbol of a new era where even the poor could afford a little of her polished luxury when the economy was pitiable. She revelled in walking in artistic circles as well, producing high quality pigments for which artists worshiped her: a brilliant yellow, the most stable of all greens and deep reds. She was also the star of the Art Deco movement, the perfect material for sculptures and decorations. This Element has claim to fame in the highly prized 'alexandrite gemstone'. The stone named after Tsar Alexander I of Russia, has a small amount of Chromium in it, but that small amount lifts it from the ordinary to the extraordinary. During the day it glimmers a blue/green but by night, in artificial light, changes to a deep red.

Soon Chromium outstretched her reach, and the public grew tired of seeing her everywhere. She became synonymous with all things tacky and in bad taste. It seemed every kitchen was covered from floor to ceiling with Chrome and hideous Chrome plated American cars were everywhere. The superficial Element's reign in the limelight was over. And by whom was she usurped in the hearts of the people? None other than the new Metal of the future; Titanium.

Manganese 25

Name: Either from the Latin 'magnes' meaning magnet or from 'magnesia nigra' from the black magnesia rock.

Subject Notes:

Subject 25 suffers from delusions of the highest order; truly believing she is on an equal footing with the notorious Oxygen. Her logic regarding this misconception is based on that she too is essential to all species. Unfortunately any similitude ends there. The amount of Manganese a human needs is modest at best and we are still unsure what her exact role is. It is clear that she is required in human metabolism, though why, we are unsure, although it is known that various enzymes in our body need her assistance to work. She also appears to be one of the most abundant Metals in the soil, another detail Manganese has assumed to make her seem more significant and indispensable in the minds of others.

Oxygen, never one to miss an opportunity, has exploited the 'ignis fatuus' of Element 25 for her own means. When Oxygen bonds with Manganese (MnO4-) she becomes toxic. Miners were often the most victimised by the duo. When they breathe in the dust they develop 'Manganese Madness'- the symptoms of which are as follows: involuntary laughing, crying, aggression, delusions and hallucinations, and lung problems accompanied by the other symptoms. In our days we encounter this very little. Thanks to Subject 25's acquired toxicity she is able to cover her weaker character flaws, such as her brittle nature, the fact that much like her neighbour Iron, she will rust in water, dissolve in acids, and once Oxygen has had her fill of entertainment from Element 25, in a powdered form, will combust under the gaze of the malevolent Oxygen.

Subject History:

It is no secret to those who have powers of observation that Element 25 loves purple, indeed Manganese is the reason amethyst is purple. This has been taken advantage of for hundreds of years by glass makers, even before we isolated her as an individual. The natural colour of glass has a greenish tinge but by adding Subject 25 it turns clear, add too much, it will turn purple.

In the Victorian era some were fooled by her posturing and Manganese became a principal in pharmacies, despite been deadly if ingested. Yet she was recommended for patients with halitosis, sore throats, and also to bathe wounds. A washing up liquid was also devised containing Manganese which had to be flavoured with lavender to stop it being mistaken for other substances. Happily these products are no longer used and her new vocation is mostly in alloys where her brittle nature is not problematic. She largely employs herself in steel (13%) adding wear resistance and workability. This alloy is used in items that need extra strength like railway tracks, prison bars, safes, helmets, and rifle barrels.

Judging by the list of her engagements her delusions do not seem to be an issue on the whole, and she has shown remarkable adaptability and resourcefulness in the face of the changing times.

25

MN

Manganese

Iron 26

Name: From the Anglo- Saxon 'iren'. The Chemical symbol 'Fe' comes from 'Ferrum' the Latin for Iron.

Subject Notes:

There always have been numerous pre-conceived ideas concerning Element 26. Ages have been named after her (the Iron Age; 1500BC, with the the Hittites), planets were assigned to her in alchemy (Mars), and even gods have been likened to her. In the Element world she is the master of industry having successfully created an enterprise that has lasted as long as humans have been able to use her – though we have no exact date for the beginning of Iron's reign we can find objects made of Iron (from meteors given their Nickel content) in Egyptian tombs as far back as 3500 BC. Today over 500 million tonnes (add recycled and it's another 300 million tonnes) of Iron are used every year. Iron has more applications than any other Metal and she accounts for 90% of all Metals refined. This has created inevitable rivalry among industrial contemporaries. It is unsurprising then that the picture created of this brawny, purposeful Element is one of a tough, unfeeling, angry mercenary of commerce, selling her war-mongering paraphernalia to the highest bidder. Our stories concerning her are little better; the Egyptians believed she came from the bones of Seth (the god of the red desert, storms, disorder, and violence), Greek legend says that two brothers murdered the third brother, burying him under a mountain where his corpse became Iron, and of course there are all the alchemy theories too which add to this aggressive picture. If we examine some of her ventures we see how Subject 26 has gained herself this reputation; with the discovery of how to smelt Iron ore (1500BC) came the predictable uses of Iron for war, murder and death on an epic scale. Swords, originally forged out of Iron, were so effective they accumulated years of bloodshed before being replaced by guns, bombs and shells which allow for a wider spread of death and destruction. Iron's ledger is red indeed.

After looking at these facts, the portrayal of Element 26 seems a sound analysis, but if we look deeper we see there is more to this Element, after all every living thing needs Iron to live. When we look at the myths that link Subject 26 to bloody gods and war, perhaps the focus should not be on the shedding of blood but on the fact blood is essentially Iron based (haemoglobin). Thus Iron gives life by delivering Oxygen to parts of the body and taking away the unwanted Carbon Dioxide.

Iron is so involved in our lives it is hardly worth listing her accomplishments as all could reel off a list of her uses. When an Element is so fundamental to our lives we have learned she usually takes with the other hand (i.e. Oxygen) but, as far as we have seen, Iron has asked for nothing. Only dangerous in excessive amounts, she is not degenerative. This speaks of a kind Element, a far cry from a merciless baron of carnage as we have been led to believe. Consider that she is easily attacked by other Elements and not impervious to the reports but is known for her rusting.

To conclude the notes on Element 26 it seems to me that Roman historian and philosopher Pliny summed it up best when he said "Iron serves as the best and the worst part of the apparatus of life."

Cobalt 27

Name: from the German 'kolald' meaning goblin. The German originated from the Greek 'cobals' meaning mine.

Subject Notes:

Cobalt is appointed in the Elemental secret service for the unique skills she posesses and the industrious methods she employs. She handles heat with interesting results and in many ways it is this quality that is the key to her success in the service. Cobalt is a ferromagnetic Element, and although her magnetic competence does not equal that of Iron, she can stay collected and maintain her magnetic prowess for a much longer period of time under heat.

Subject 27 is also the Element responsible for 'synthetic ink' or 'invisible ink'. A message written in her ink stays imperceptible until heated. Despite her puckish, dark humour, and enthusiasm for quips of a disagreeable nature, this 'trick' ensures her high rank in the secret service. Her inks were first employed in the 17th century to send letters of a sensitive nature or to aid espionage. Later in the 19th century the same trick was adopted in a less shady procedure in artificial flowers, popular in the hats and homes of ladies of fine breeding. Petals would be dyed pink but when the weather became warm and sunny their colour would change to violet and blue hues depending on the heat of the day.

With further investigation one can see this is not the only instance where Cobalt's stunning colour has brought her wealth and exaltation. She has been used from as far back as the ancient world as a pigment to dye or paint objects vivid blues. Evidence of her presence can be seen from Chinese pottery and glazes, to artefacts in Tutankhamen's tomb. She also enchanted Venetian glass makers; it was indeed her rare ores and colour that brought them much acclaim.

Most life requires her as she is the main player in vitamin B12. Without her the body could not produce enough red blood cells and would be starved of Oxygen. B12 also maintains nerve and other tissues. A lack of Cobalt will result in pernicious anaemia. However not all of Element 27's press is of a reputable bearing. As suggested earlier she possesses a dark sense of humour and enjoys playing with the minds of others. This characteristic has given her a reputation of a rather unsavoury nature and makes many Elements wary and uncomfortable in her egotistical presence.

In the 16th century she and Arsenic teamed up to play a most insidious jape. Miners in Germany attempted to smelt what they thought was Silver, however the substance was much more sinister. The matter was in fact 'smaltite', our duo in question: Cobalt and Arsenic ($CoAs_2$) posing as Silver. The miners soon became ill on the toxic fumes (Cobalt is actually of low toxicity, so the offending Element was in fact Arsenic). The miners began to curse the substance, claiming the mine had been bewitched by goblins. It is this tomfoolery that led to Subject 27 receiving her name. Nonetheless Cobalt is a hard working Element and puts her craft and work ahead of anything else. As we have seen she is actually a relatively non-toxic Element and most of her work is for the good of the human and Elemental worlds, with only the occasional deceit.

Co

27

Cobalt

Nickel 28

Name: Shortened from the German 'kupfernickel' meaning 'Devil's Copper' or 'St Nicholas's Copper' – 'old Nick'.

Subject Notes:

Many of the Transition Metals are wholesome Elements, working hard, and endeavouring to improve public life for everyone irrespective of class. Nickel, however, is of a darker calibre, with a deviant mind and corrupt morals. That is not to say she does not work hard, for she has always been at the forefront of innovation, motivating other Elements and adding her strengths to theirs to create highly resistant alloys. Never-the-less what Nickel gives she will also take away.

It is suffice to say Subject 28 brings many qualities to her profession, she is ductile, malleable, a good conductor, and like her sisters Cobalt and Iron she is also Ferro magnetic. Perhaps the greatest characteristic she brings to industry is her power to resist. This formidable Element has a high tolerance to the wiles of Oxygen and can resist corrosion and Oxidisation; this makes her the perfect business partner for Iron, who will easily degenerate under Oxygen's malcontent. Together they make 'Stainless Steel', as strong as Iron but with the defiance and malleability of Nickel. Just one example to demonstrate the business capabilities of the ingenuous of Subject 28.

The price of her devilish charm and labour is often high. On examination one observes several risks to health. 10% - 20% of people on contact with Nickel will get dermatitis, workers who breathe in Nickel dust develop bronchitis, lung and nasal cancers. The most diabolical of her praxis is Nickel Carbonyl gas, which, if inhaled in any amount, can kill. Thus the darker, sinister side of her 'industry' is revealed.

Subject History:

Nickel, being a key business partner of Iron, spends large amounts of time in her company. Evidence of their coalition can be found long before the intervention of humans, in meteorites and in the earth's core. Much to the chagrin of German miners her affiliations have also spread to other Elements like Copper and Sulphur. The miners believed little devils or goblins stopped them extracting Copper from Nickel Arsenide. This story, like that of Cobalt, is how Subject 28 came about her apt name. Nickel has been at work for many thousands of years: swords have been found made of Iron and Nickel in an early form of stainless steel, green and blue beads have been found in Egyptian tombs, the Chinese mistook her for a form of white Copper and the Peruvians mistook her for Silver. She has been used in coins, and to some extent still is. She could have made it to the high rank of 'Coinage Metal', even having the American 'Nickel' named after her. However this honour has eluded her on account of her viperous, un-lady like behaviour and her toxic schemes.

Most of her talent is taken up by 'stainless steel' in everything from jewellery and watchstraps to spectacle frames (some who are sensitive to Nickel get 'Nickel Itch' from 'Stainless Steel' even more so as sweat begins to dissolve the Nickel). Despite her high price and devilish streak her properties make her too useful to abandon completely.

Ni 28

Nickel

Zinc 30

Name: From the old German word 'zinke'. Meaning pointed or tooth; a reference to the sharp tooth like crystals formed after the smelting process.

Subject Notes:

Subject 30 is shrouded in symbolism and sorrow. Zinc, a demure Element, is physically similar to Magnesium, and as a result is very valuable to humans and the rest of the natural world. A lack of Zinc causes growth retardation in children (Zinc deficiency is responsible for the deaths of approx. 800,000 children worldwide per year), through infection susceptibility, diarrhoea etc. While we need her biologically her main vocation is more tragic for the brittle yet noble Element: she is a sacrificial anode, a martyr to the cause of other Elements. She ransoms herself as a coating on other Metals putting aside personal interests and well-being so that other Elements will not be eroded by salt water or the insidious Oxygen. For this willing victim more than 50% of her work load goes into galvanizing other alloys, so that objects like bridges, boats, motorway guards and roof tracks last longer.

Zinc is set apart from others as her status is a chalcophile Element. This means she does not succumb easily to the guiles of Oxygen. This is not to say she is without weakness, she readily complies and combines with Sulphur or others with the chalcogen 'gene'. Though the unhallowed unions with other Elements may cause her to lose her shine and become tarnished they also provide her with a carbonate layer ($Zn5 (OH)6 (CO3)2$). These coalitions help her with her crusades, making her a stronger Element in the face of corroding influences. These chalcophile Elements also have another characteristic, they form compounds which do not sink readily into the Earth's core and remain on or close to the surface. It is these Elements that usually have long histories with humans and that could be easily found.

Subject History:

Zinc has been used since the 13th century BC, longer still if one counts her part in brass (Zinc and Copper) in which she has been used unwittingly long before Zinc herself was discovered. Her name has popped up in many places over the centuries from Judea to Greece, from China to India: Which is where Subject 30's story starts, Rajasthan; India is one of the oldest places to use Zinc on a large scale in her own right. In fact she is the only Element with a determinable discovery date where western science cannot claim credit for her discovery.

Zinc's story is disconsolate, perhaps fitting considering her scapegoat position. Zinc first arrived in Europe via China. At a time when Alchemists had already assigned planets to Elements and given them partners in their pseudoscientific world, she was in 'no man's land' not a Metal of the ancients nor of the modern world. Though alchemists enjoyed burning her and gave her various names like 'Philosopher's wool' or cruelly 'mock Silver' she was never fully accepted on the level of Iron, Gold or even Antimony. There was no great celebration for her entrance; she was simply put to work in the galvanizing process, and as pure white pigment in papers, paints and cosmetics.

Yttrium 39

Name: Named after the village Yitterby in Sweden.

Subject Notes:

Many Elements are unhappy about the lot they have been given, whether it is work of a tedious nature or the lack of eminence given to them, but few protest about the 'group' in which they find themselves. Yttrium however is deeply perturbed about having being named a Transition Metal and views herself as a Lanthanide (a group of Elements researched in Volume iii). Indeed she has much in common with the Lanthanide Elements and one will almost always find her associating with her friends of that group rather than those who surround her in the Periodic Table (Scandium and Zirconium). The Lanthanides are also called 'the Rare Earth Elements' in some circles, and Subject 39 was the first Element to be classed as a 'Rare Earth Element'. This speaks of a close bond, but it is more complex than a simple preference. Element 39 has the same biology or atomic radius as the Lanthanides making her chemically more like them than she is to her surrounding Elements (as one would normally expect in the Periodic Table). Because of this she acts as if she was one of her beloved Lanthanides.

Due to this perceived slight regarding positioning and her Lanthanide genetics she is more than a touch unbalanced and aggressive. She will burn arduously if ignited and is attacked by water. Measures have been taken that seem effective in keeping Element 39 comparatively sane and calm, though, do nothing for her irritability. More than any other Element Yttrium shows a high affinity with Oxygen, therefore in a manner of speaking Oxygen has become her keeper; in an uncharacteristic act of charity Oxygen keeps Subject 39 in line and helps protect her from further misfortune that may set off another outbreak, by creating an Oxide layer that helps prevent reactions.

Subject History:

Yttrium has the ability, if she chooses, to improve the work of other Metals, for instance she renders Chromium able to produce a finer grain and if added to Iron makes her more workable. Evidently her dislike for being classed as a Transition Metal does not extend to hatred of the Elements themselves and she is still willing to work with them.

The most important toil of Subject 39 is in making 'phosphors'; such as the red ones used in cathode ray tube (CRT) displays and in LEDs. She is also used in the production of electrodes, lasers and superconductors that can conduct electricity without losing energy and in making camera lenses shock resistant.

Y

Yttrium

Zirconium 40

Name: From the Persian word 'zargun' meaning gold-like.

Subject Notes:

This hard, lustrous, yet brittle Element attempts to defy her working class roots and tries to direct her employments towards jobs of a more feminine, refined or artistic nature where possible. Holding "her Majesty" Gold in high esteem she endeavours to have the same elegance and refinement as her sovereign. Unlike many Elements who long for fame, Zirconium is not bitter and pines after the glamour with good grace and a sense of humour, making her a more palatable Element.

Element 40 seems to be a creature in need of constant social interaction and is never found alone. Usually found in the company of her sister Element and best friend Hafnium, it is almost impossible to tell the difference between the two, who spend so much time together they have taken on each other's traits and mannerisms. Consequently it has become almost futile to attempt to separate them. When it comes to employment because of her group instinct, Zirconium is easy to work with and is never found shirking her work load.

Subject History:

Element 40 strives for employments worthy of the most glamorous and charismatic Elements and to some degree she has succeeded. Her larger crystals are sold to the public as semi-precious stones in a selection of colours. Her most sought after and popular product is her gold coloured stones. Once cut, cleaned and polished so impressive are they that some have mistaken them for diamonds – indeed her gems were known in the Bible times.

Other vocations Subject 40 deems desirable and aspirational include her roles in the cosmetic industry, anti-perspirant, 'fake' gems, and her use in pigments (she creates yellows and blues). At one time she was used in a cream to treat poison ivy rash, although in 1960 this was discontinued as it also cause nasty reactions.

However for an Element of Zirconium's talents it cannot always be glittering gems and itch inducing creams, she also must work in other, what she considers, 'dull jobs'. Thanks to her heat and corrosion resistant qualities she is used in the lining of furnaces, in giant ladles handling molten Metals, foundry moulds, glass and ceramics. Her work in ceramics has led to opportunities helping her to make progress toward her dream. Known for her reliable and even finish in glazes, she has become more popular in this area and demand for her has risen allowing her to spend almost half her efforts working in the glazing industry.

Subject 40 has other talents (which she tried to hide so as to avoid work in the chemical and nuclear industries) which were uncovered around the 1940's. Zirconium is ideal for nuclear reactors as she does not corrode, can deal with high temperatures and does not absorb neutrons from radioactive isotopes. As a result, and much to the Element's dismay, this industry tries to buy all of her Metal. Fortuitously for her, her friend Hafnium is better equipped for this job and tries to help with the work load (see Hafnium) in an attempt to free up Zirconium's time for more desirable pursuits.

Cadmium 48

Name: From the Latin word for the mineral Calamine - 'cadmia'. Some believe she was named after the Greek 'Cadmus' in mythology, who founded the city of Thebes.

Subject Notes:

The ill-natured Cadmium is the Elemental world's artist. Notorious for her inspirational art, extraordinary paints, and unhappily, her eccentricities, foul moods, and anti-social behaviour. She is tolerated by others solely because of her artistic genius. When Cadmium emerges from her studio the results can be devastating. Subject 48 is like Zinc in biology and thus is freely admitted into the body where she evicts Zinc and Calcium. Once there she does not possess the skills to complete the operations Zinc has in place, and begins to cause damage. Exposure to Cadmium fumes causes chills, fever, and muscle ache often referred to as "The Cadmium Blues." If there is no respiratory damage these symptoms may resolve themselves, but more exposure could cause tracheo-bronchitis, pneumonitis, or pulmonary oedema. If consumed she can be fatal. Ingestion of Cadmium causes immediate poisoning and damage to the liver and kidneys. The bones become soft and weak, causing pain in the joints, victims become so frail their body weight is enough to cause their bones to fracture. Under her borderline psychotic behaviour and her egotistical rants we have discovered an over-sensitive and deeply nervous Element. Her constitution is soft and easily cut with a knife. When out in the public eye she becomes dull and tarnished (Curiously a contrast to what the bright and garish paints would suggest), signifying that her rakish demeanour and excellent works of art may be a front to hide the real Element behind, never letting anyone close to the real Cadmium. This is however pure speculation, as I seek to find a motive behind Element 48's actions.

Subject History:

There was a time when Cadmium was used in everything. On realising her toxic nature, the government tried to ban her. Artists universally were outraged, for they had fallen in love with her vibrant paints, much more brilliant than other muddy coloured pigments. This however was not the only issue artists had; it was also what the ban would imply. They explained that Cadmium was only dangerous when it decayed into the ground. They argued that their works of art were of such outstanding skill and quality they would never be thrown out to decay. Therefore the pigment was perfectly safe, the government agreed and Element 48 had been saved by artists. Another key event in Subject 48's history worthy of note happened in a town about 200 miles North West of Tokyo. The villagers started to become very sick, as unknown to them a nearby factory was disposing Cadmium upstream. Their fields were polluted by the Metal. The rice grown in Cadmium contaminated soils contained more than ten times the Cadmium content of normal rice. Not knowing where their condition came from they called the 'disease' Itai-Itai which means ouch-ouch. The excess Cadmium interfered with Calcium deposition in their bones. The 'ouch-ouch' factor of this disease was the result of weakening bones liable to collapse. The industry responsible for this were eventually forced to take responsibility for their actions.

48

CD

Cadmium

Hafnium 72

Name: From the Latin name for Copenhagen – 'Hafnia' – the place of her discovery.

Subject Notes:

Hafnium is the quieter and more reserved of the two inseparable friends (Hafnium and Zirconium), and follows timorously in Zirconium's wake of glitter and farfetched fantasies. Subject 72 does not seem overly taxed by living in Zirconium's shadow and is happy to work behind the scenes to help her friend advance in any way she can. The unobtrusive Element was so well hidden by Zirconium's outgoing personality and ambitious projects that Hafnium was the penultimate nonradioactive Element to be discovered. The foremost reason for this is that Hafnium's chemistry is so similar to that of Zirconium and their atoms are the same size (rather than getting bigger as is the usual case in the Periodic Table for columns going down). It is all too easy to get the two mixed up, especially when they spend so much time together.

Subject 72 is however just as capable as Zirconium in most things, being just as corrosion resistant, ductile and having a high melting point. In the case notes for Zirconium there is reference to Hafnium being better equipped for work in nuclear reactors. In fact Element 72 is five hundred times more effective at absorbing neutrons than her friend, in an attempt to save her friend from the 'dull jobs'. The only issue with this is that only fifty tonnes of Hafnium is refined each year, making her an expensive Element to employ. It is not that Subject 72 is a rare Element, it is rather her closeness to Zirconium that makes her hard to separate, but once alone she excels in high temperatures, alloys and ceramics.

Subject History:

The discovery of Hafnium has been a trial for reasons explained above. The greatest scientific minds of the time knew there was an elusive 'missing Element' that lived under Zirconium in the Periodic Table, it was only a question of finding her.

In 1911 Georges Urbain (1872 – 1938) believed he had finally found this modest hidden Element, naming her Celtium. Sadly it was a case of mistaken identity.

The young scientific genius of his age Henry Mosely (1887 –1915) discovered a technique in which from the X-rays Spector of Elements, he could deduce that Elements 43, 61, and 57 were still missing. Hafnium would also have been on the list of 'missing Elements' had it not been for the recently announced 'Celtium'. Sadly Mosely was killed at the tender age of 27 by a bullet to the head in the World War 1 Dardanelle's campaign. He would never see the discovery of Hafnium.

Hafnium could not stay hidden forever, it was only a matter of time before Celtium was discredited and the race for Element 72 was back on. Eventually in 1923 Dirk Coster (1889 –1950) and a Hungarian chemist by the name of George Charles de Hevesy (1885 -1966) found the Element in question hiding under the metaphorical skirts of Zirconium - in her ores. She was then named after the City in which she was finally uncovered rather than the scientist's homes, in an act of modesty befitting Hafnium.

72

Hf

Hafnium

Mercury 80

Name: Named after the planet Mercury. The Chemical Symbol comes from the Greek word 'Hydrorgyrum' meaning 'Liquid Silver' also known as 'Quick Silver'

Subject Notes:

One of the most notorious Elements throughout history, Subject 80 has earned herself a reputation for insanity, outlandishness and murder. On dissecting her personality we find that she is something extraordinary, with her lustrous glimmer and liquid movements. She is one of the few Metals that is liquid at room temperature and has captured the hearts of Emperors, Alchemists and Artists alike. Yet once we strip away the myths, mystical rituals and fantastical tales created over hundreds of years of alchemy, we find that she's actually a rather unimpressive Element with little to offer the industrial world other than madness and a trail of disappointment and chaos.
To better understand this Element and her legacy it is beneficial to study her history, the rise of her infamy and the fall from grace on discovery.

Subject History:

Although she has been used world wide we see a particular love of her in ancient China, where Emperors believed she was the elixir of life - they drunk Subject 80, ignorant of her toxic effects. In England there is evidence of her being used to tell the future by druids. In Rome she was associated with Mercury the god of commerce, travel and thieves. She was key in alchemy and in Africa it is still the practice in some cultures to use her to ward off evil spirits. An enquiring mind may ask how she came by such a mystic and reverent status? The answer lies in alchemy. Mercury was believed to transcend the liquid and solid states. This belief carried over into other areas. She was thought also to be able to transcend life and death, heaven and earth. As time progressed so did human learning. Unfortunately we still didn't make the connection between Subject 80 and her toxic properties and the Victorians fell in love with the fabulous psychopath all over again. She was employed in dentistry, a 'cure all' pill, perfumes, lotions, soaps, make up, a vibrant vermillion red paint and of course made famous by Lewis Carroll's 'Alice in Wonderland'; The hatter becoming insane after a long period of time in the company of Mercury- this was not limited to fiction. In 1693 sir Isaac Newton had had an unexplicable breakdown after dabbling in alchemy. His symptoms resembled that of Mercury poisoning. Even the hard-of-cash king Charles II was a known alchemist and built a lab in his basement, his cheery disposition suddenly turned into irritability. After a spillage of Element 80 he became slurred in his speech, then suffered fits and convulsions, it was not long before he died. Other symptoms of this Element include salivation, bad breath, inflamed lips and gums, vomiting, respiratory distress, kidney damage and tremors. She attacks the nervous system (a tell-tale sign is 'spidery' hand writing), then she causes mood swings from timidity to irritability, out bursts of anger then apathy, and depression to paranoia. Thus we can clearly see reasons for the 'hatters' madness. Eventually we came to our senses and Mercury's products were discontinued. Today she is limited to specialist paints, older thermometers and relics of a bygone era.

Coming Soon!

The Illustrated Guide to the Elements

Volume III
The Final Episode

The Lanthanoids, Actinoids and the Mysterious Synthetic Elements!

Sub-Chapter
The Coinage Metals

It is at this conjuncture we investigate the first of our Transition Metal sub-groups: The Coinage Metals. As the name implies the Elements in this chapter have historically been used in mint coins, or as the prominent Metal in a coin alloy.

These Elements are the cream of the crop. Only the most important and competent Elements are chosen for this title and honour. This elite group has strict criteria to meet and very few others ever enter their exclusive inner circle. They must be 'wear resistant' and possess anti-corrosion properties for a start; other required qualities will be discussed in subsequent 'Element case files'.

This group can not be perfectly defined because there have been many Metals and alloys used in coins over the ages. Some Metals have been used too briefly to be worth a mention in this chapter, others, their parts are too small to qualify. While some Metals have been used as coins but have never been monetized in any nation or state, these have been called 'Demonstration Coins' and will also not be included in this section. While examining texts with regard to this group it is apparent that a select few are truly deserving of a place in this study. A small handful who have turned up time and again, revered and constant enough in their use as coins to truly be considered Coinage Metals. Three metals to be exact fit this bill and thus find themselves propelled into their superior positions and this chapter: Gold, Silver and Copper.

Copper 29

Name: Old English name 'coper' – which originally came from the Latin 'sypriumaes' meaning 'from Cyprus'. Chemical symbol comes from the Latin.

Subject Notes:

Subject 29 is a very capable, stable and ruthless Element who moves in circles of the highest pedigree. She is in possession of a long canon of qualities which have made her invaluable and influential in both the Elemental and biological worlds. She is malleable, ductile, a sterilizing agent, a good conductor of both heat and electricity (second only to Silver), as well as being resistant to corrosion in water and air. All facets she exploits to the highest degree in the forwarding of her industrial domain. This highbrow Element shares a trait that only one other Metal possesses - Ms Copper is a coloured Metal, having a warm reddish hue. It is this that has given her the opportunity to climb the social ladder and rub shoulders with the most exalted of all Elements: Gold.

This Element enjoys the eminence and glory she receives, for both her high social status and the impeccable and diligent way she completes her work. Copper also takes great pride in the fact that her work is not only of the highest quality but also aesthetically pleasing and affordable.

Subject History:

Copper has an enduring and enthralling past. In ancient times she was used in everything from jewellery to tools, from weapons to cutlery. From this she slowly started building her empire socially and economically.

Cyprus has been the main exporter of Copper since before the times of The Roman Empire. Investigation has uncovered a rather surprising fact; so great was her fame in Cyprus that the island was named after her rather than the reverse.

All life needs Element 29 to survive; in fact some animals have her in their blood rather than Iron. As with the majority of Elements I have investigated, an excess of this Element can kill. In the case of Copper however one usually vomits before any permanent damage can be done.

This self-confident Metal has sterilizing properties that make her perfect for coins and pipes, and this holds the key to her fame. When a germ crosses her it begins to absorb Copper atoms. This disrupts its metabolism, thus killing it in a matter of hours. Silver also possesses this quality and for hundreds of years was used to clean water, cleanse wounds and later to conduct electricity. Copper, having no guilt, took over Silver's enterprises, and the public embraced her much cheaper services. This is evidence as to just how remorseless this Element can be when she can provide a more efficient and more effective service - proving there can be no sentiment on the road to industry and improvement.

CU 29

Copper

Silver 47

Name: Sanskrit for white. The symbol 'Ag' comes from the Latin word for Silver 'argentum' meaning white or grey.

Subject Notes:

For thousands of years Element 47 has been steeped in symbolic meanings which reveal much about her nature and personality. Such symbols include: Feminity, hope, innocence, and most often purity. This elegant and mild Element is overly critical of herself, obsessing over her flaws and faults, striving to maintain her purity and virtuous image. Unfortunately piety comes with the inevitable fall from grace. When exposed to the world for too long she becomes tarnished or blackened, and the only way to rid herself of this black layer is to remove it. This leaves Subject 47 with a slight air of melancholy and a wistful glaze, for the price of purity and a well sought after shine is to lose a small piece of herself with every exposure. This leads Silver to works of charity for the less well-off and a personal crusade to better herself. Despite being of higher class, she can never truly know how the 'other half' live, limiting both her success and helpfulness.
Silver is the best conductor of electricity and heat of any Metal, as well as being a sterilizing agent (as alluded to in Copper's case notes). However being of a noble upbringing her services do not come cheap, and so we often find she is replaced by other more economic Elements.

Subject History:

Silver has always been held in high regard by both the prosperous and the underprivileged. Element 47 has been used for medical applications for many centuries due to her purifying and sterilizing virtues. Silver is able to kill bacteria in external wounds leading physicians to use dressings impregnated by her. Subject 47 may have lost out in some vocations but her use in medical wound dressings is of increasing importance due to the recent surge of antibiotic-resistant bacteria.
Phoenicians stored water, wine and other liquids in Silver jars to prevent spoiling and even as recently as the 1990's silver coins were deposited in milk bottles for the same reason.
In the same vein some take her as a supplement, as a cure all, though governments warn against this course of action as benefits have never been proved scientifically. Be that as it may, being too fond of this Element does have its consequences. Ingesting too much Silver may cause 'Argyria', a side effect of which causes the skin to permanently take on a grey-blue appearance. This historically occurred in the more opulent classes who often ate and drank from Silver utensils or jugs while showcasing their wealth, from where the term 'Blue Blood' originates.
Despite losing fractions of her trade to other Elements, in some areas Silver is one of the few Elements whose demand is as great today as it was thousands of years ago, - for medical uses, jewellery, trinkets, utensils and cutlery. Evidence suggests that trend will not end in the near future, and her prominence and wealthy lifestyle is safe for a good while longer.

AG ⁴⁷

Silver

Gold 79

Name: Anglo-Saxon word, which may have come from the word 'geolo' meaning yellow. The chemical symbol originates from the Latin 'aurum' meaning glow.

Subject Notes:

Gold, the most sought after and revered of all Elements. She is not the rarest or most expensive of Metals but even after all these centuries she is still the most desirable. Subject 79 rules over the Elemental world and like many of the ruling classes is relatively useless and overrated. That is not to say she does not have skills. She is the most ductile and malleable of all Metals and can be easily shaped or beaten paper thin. On top of this she is largely incorruptible, unaffected by the perils of the lower classes, impervious to water, air, acids and alkaline (though she can be dissolved in Aqua Reiga, but can easily be regained none the worse for wear).

Eager to hold on to her electron wealth she has a large positive pull (79+) keeping her electrons close, unwilling to share or let them out of her sight. As a result she is rarely found mingling with the commoners in compounds. If such an event occurs Gold is easily extracted, all too willing to leave their plebeian company. To ensure all know of her status she is one of only two coloured Metals (Copper being the other) and garishly glitters her gaudy yellow, unhindered by ores.

Subject History:

Gold was one of the Elements first known to us, largely because of her laziness and anti-social royal attitude. She could be found on river beds unhindered, sparkling brilliantly to announce her regal presence. This is when the human love affair with the warm glowing Metal began. Since then she has been used in jewellery, tombs, decorations, golden leaf, electronics, in an aid to help arthritis sufferers, and in paints. Shown off by the rich, coveted by the poor and the ultimate goal of the alchemist.

It is not just her incandescence and aura that have captured the human heart. She has long been associated with immortality, due to her relationship with fire and her purity. Though fire may melt her down, she loses nothing from the heat and when cooled has the same properties and appearance as prior to melting. Her incorruptibility in the face of Oxidization and corrosion has given her a place of high moral standing among Elements and people alike, causing them to lend her great name to the 'Golden Rule'.

No matter how many new Elements are discovered, how useful or rare they are, it seems none will replace Gold. After thousands of years she is just as important and valuable as she has ever been. Even when money loses value, Gold remains a commodity. Thus it looks like Subject 79 is set to rule the Elemental world indefinitely, despite the constant attempts of others to usurp her.

79 AU

Gold

Sub-Chapter
The Refractory Metals

The Refractory Metals are a class with high ideals and high melting points, further-more they will go above and beyond the average Element. They have successfully created a union type structure to support each other in their struggles and advance the excellence of their work.

As we peer deeper into the infrastructure of this party we see why, though many Ele-ments may agree with their manifesto, few can live up to their massive workload and stringent principles.

Their Manifesto explains that the bourgeois Elements exploit the proletarian Ele-ments (like themselves), with low electron wealth, so they may accumulate the wealth for themselves. Meanwhile the 'lesser Elements' will fight amongst themselves and the competition amongst the working class will take the attention away from the affluent lifestyles of the rich, thusly keeping the lower classes low and the rich rich. They fight against this system, which in their view, rests entirely on the antagonism among the workers, and believe that through hard work and pulling together they will eventually rise to power through a class struggle.

This interpretation of the Element class system has created enemies, admirers, and want-to-be Elements, but opinions are of no consequence to them - only the work. As part of their declaration there is a list of capabilities that true workers and revolu-tionary Elements of the Refractory Metals should exhibit;

- Refractory metals are a class of Metals that are extraordinarily resistant to heat and must at least have a Melting point above 2000°C

- They should have great toughness at room temperature as well as being resist-ant to the wear and tear of everyday life and extreme conditions.

- It is requisite that they be chemically inert so as not to give in to the tempta-tions of electron rivalry or the bourgeois debauchery.

- A Refractory Metal is stable against creep deformation to very high tempera-tures. (Partly due to their high melting points)

- Though not mandatory it is respectable to have a relatively high density.

These are the main qualifications for the committed Refractory group. The definition of which Elements can belong to this group differs, but the Elements I have uncov-ered that seem undeniably part of this class are as follows: Niobium, Molybdenum, Tantalum, Tungsten, and Rhenium. The Refractory Metals are of three distinct and different backgrounds in the Periodic Table, so there is a wide variety of chemical properties they individually possess not listed above but which will be disclosed in their individual case files.

Niobium 41

Name: Named after Niobe from Greek mythology, the daughter of King Tantalus, goddess of grief – because of her similarities to Tantalum.

Subject Notes:

Usually found with her comrade Tantalum, Niobium is the 'Elements Element'. Her mission to convert other Elements over to the Refractory revolution is realised by setting an example of hard work, cooperation and unremitting adherence to their ideals. Her objective is to improve 'the work'; she strengthens and inspires the workforce, consequently she is well liked throughout the Element world. Fortuitously for us her commitment to recruiting others and rousing them is good also for our world. Indeed it is hard not to like this non-toxic Metal, particularly when drawn to her specifically formulated iridescent radiance. Like many, Niobium utilizes an Oxide layer to protect herself from corrosion and corruption, but if she lets the coating get thick enough it will begin to reflect light. This reflected interference pattern gives the appearance of shining multicolours. Depending on variables like the angle of one's view or the thickness of the layer, different colours are created. It is said Subject 41's purples and blues are particularly disarming – much to her disgust this is becoming more often used in jewellery. Niobium will use any technique at her disposal to draw others into the Revolution whether it is her striking smile, impressive commitment to drudgery or her aptitude for enhancing others labour.

Subject History:

Charles Hatchett first isolated this Element in1801. Despite her impressive capabilities, work ethic and considerable contribution to our lives Subject 41 is not a well-known Element, not through any shyness or false modesty, rather through her sticking to the philosophy of the Refractory Metals. i.e. wishing for the attention to be given to the revolutionary Group rather than chasing personal acclaim or fame. Only a small input from Niobium is needed to inspire prodigious results from other Metals. Her involvement imparts great strength and enthusiasm to almost every other Metal. Alloys with Nickel, Iron, and other well-known Elements flourish. Examples of her associations are worth examining: in 'Stainless Steel' so small is her part that many do not know of her involvement, yet she assists in it's ability to be welded and adds workability. When working with Titanium they become superconductors (though they work best at -250̊C). Other Oxide Metals lose superconductivity as the electrical current is increased this union allows superconductivity even with a high current density. They have revolutionized the construction of high-field strength magnets for research purposes (MRI).

Her Refractory Metal properties can supplement Elemental exertion with the lightest of touches, without adding disproportionately to the weight, and consequently alloys containing Niobium are used in aircraft turbines and aerospace rocket engines.

41

NB

Niobium

Molybdenum 42

Name: Derived from the Greek 'molybdos' meaning Lead.

Subject Notes:

Over the years Subject 42 has been mistaken for many other Elements, for instance Lead and Carbon (graphite), but with the constitution of heroes and being an Element of calibre she has eventually won herself a reputation and name of her own. Her Element in arms and fellow revolutionary, Tungsten may be in charge of the Elemental military as General, but Molybdenum is her close second as a stalwart Colonel and weapons expert.

She has an array of qualities that make her perfect for the design and modification of weapons: resistance to corrosion and corruption. She has little time for the machinations of Oxygen and is repelled by her fabrications and bourgeois tendencies.

Subject 42 is a formidable Element having an exceedingly high melting point (2,623 °C). As a result she can only be used in a powdered form by us and other ways have had to be devised to make her cooperate with our needs; as high temperatures are out of the question, high pressures have to be introduced.

With incentive and the leadership of her ranking officer Tungsten, to whom she shows a great admiration, and fervent loyalty to 'the cause' she has shown herself to be a hard and dependable soldier and worker.

Subject History:

Molybdenum is used in glass furnaces as she is impervious to high temperatures and the molten glass does not affect her sturdy constitution. She is vital to all species, and often used in electronics. But the vocation for which she is most accomplished, and suited is in weaponry. As we look at her résumé over hundreds of years we see the varied enhancements she offers to the industry of death.

In our quest to find the particulars of this Element we must first look back to 14th Century Japan and inspect the highly prized katan swords of the samurai class, renowned for their strength, beauty and malleability. They were superior to any other weaponry of their time and they owe much of this to Element 42, indeed, this was the first instance of 'Moly Steel' which would not be discovered or used elsewhere until the 19th century; though many of the secrets regarding discovery and techniques employed died with the sword smiths of the era.

Few other Elements can give the expertise and efficiency that industry and the military necessitate: high strength and resistance to corrosion, good conductivity skills and low expansion when exposed to heat (Subject 42 has the lowest expansion rate of all Metals). It is no wonder then that in World War 1 the master behind 'Molybdenum Steel' or 'Moly Steel' was used to plate the steel on tanks. This gave them better protection, and made the tank lighter. The result? The thickness of the steel used could be reduced allowing the tank more speed and manoeuvrability without compromising the hull strength and therefore the lives of the soldiers. This enactive and cunning was then able to be used beyond the art of death in things like aircrafts, cars, engines and rockets.

42

MO

Molybdenum

Tantalum 73

Name: From the Greek Legend 'Tantalus' King of Sisyphus in Lydia who stole food from the Gods and became immortal but was punished by being forever hungry and thirsty.

Subject Notes:

Tantalum has many commendable qualities; she is resistant to corrosion, strong and unwavering in her beliefs. However she does not share the charisma and charm of her close compatriot Niobium and can be difficult, judgmental and obstinate. She holds firm in the belief that Elements not committed to the cause are not worthy. She has all the aptitudes expected of a Refractory Metal and has earned her place in the radical group: she is ductile, hard, dense, easily fabricated, and highly conductive. Unaffected by almost all the usual chemical attacks that cause Elements distress she has only two weaknesses: Hydrofluoric acid and Molten Alkalis.

As mentioned Subject 73 can be a judgmental individual and stubborn. If she likes an Element she is one of the most generous and supportive cohorts to have, enhancing their abilities, making them super strong and lending them her powers of protection. One of the finest examples of this kind of collaboration is when she is in league with Carbon (TaC Tantalum Carbide). Together they form a substance that is harder than diamond, with a melting point of 3738C which is put to good use in special cutting tools. These alliances are so effective they are given jobs only the toughest would survive like turbine blades, rocket nozzles, and heat shields. However Tantalum does not find all Elements as pleasing as Carbon. Once she has decided an Element is un-deserving or clings too hard to a bourgeois lifestyle she actively works against them, making them weak and brittle, showing just how tenacious a character she can be. The chemical inertness and calm temperament of Tantalum makes her a valuable comrade for laboratory equipment and her toughness makes her perfect for surgical equipment.

Subject History:

Her name gives away much about the character of this sturdy Element, or at least the appetite that scientists had to find her. To discover more about the particulars of subject 73 one must enquire deeper into the fable from which she got her name. The cruel and petty Greek Gods punished Tantalus for taking their food by making him eternally hungry and thirsty in his newly gained immortality. He was made to stand in water that recedes whenever he tried to drink and near fruit vines always just out of reach. This is one of the reasons Element 73 received her name, an Element always just out of reach in her awkwardness, tantalizing her pursuers. The year was 1802, Anders Gustav Ekeberg declared he had found a new Metal. However contro-versy erupted when British chemist Wollaston said she was nothing more than her close sister Niobium, which suited the obstinate Tantalum, happy to work in anonym-ity. It was not until 40 years later in 1846 when Heinrich Rose managed to separate the two Elements, proving they were different entities. Tantalum's difficult nature prevailed and it was not until 1903 that an appearance by the Element was seen.

T

Ta 73

Tantalum

Tungsten 74

Name: The Chemical symbol 'W' from the German 'wolfram'- which in turn comes from the mineral wolframite. Wolframite means "the devourer of tin" since the mineral interferes with the smelting of tin. Tungsten from the Swedish word 'Tung-sten' meaning 'heavy stone'.

Subject Notes:

Tungsten; the ultimate Metal of war and General of the Element military forces. She is a natural leader of Elements and has earned her place as General not only through her contributions to armour, weaponry and tactical significance but also by holding other Elements together and helping them through stresses like heat. Element 74 is able to inspire others and help them keep their composure. This is mostly due to her sturdy structure and long list of accomplishments. Some of her awards and achievements are as follows; of all Metals (in pure form) Tungsten has the highest melting point, the highest tensile strength, lowest vapour pressure and lowest coefficient of thermal expansion. These specialist capabilities originate from the heavy-duty covalent bonds formed between her atoms by the 5d electrons. This brave, confident and dense Element has proven herself to be cool under pressure and instrumental to the war effort, though if contaminated with other materials, Tungsten becomes brittle and difficult to work with.

Subject History:

Subject 74 is possibly best known for her part in light bulb filaments and drill tips, but in this inquest we will be concentrating on her part in war, armaments and armour as that is her main vocation in the Element world and has a fascinating history in which her wolfish nature is revealed.

A superior steel is created by the input of Tungsten. She makes it stronger and harder, also allowing the Elements involved to withstand extreme heat without deforming or crumbling in the face of attack. Naturally this 'self- hardening Steel' was employed in the war effort, in guns and cannons. The Germans soon perfected the art and found with the help of Tungsten their guns had a longer working life and could fire 15000 rounds safely, which other countries' artilleries couldn't even touch. As the news of Tungsten's remarkable capacities and leadership qualities spread the demand for her services did also and a shortage soon hit Europe.

The fact she is found in the mineral wolframite should have given us a clue that Element 74 is capable of occasional duplicity and underhand dealings. In World War 2 the Spanish were the main purveyors of this Element and all involved in the war where hungry to employ her services. Rather than limit her options to one side the Spanish sold to both the Germans and English claiming loyalty to both while dealing under the table to the other, thus Element 74 lives up to her wolfish name.

With her triumph on the battle field and talent for enhancing others she has been able to recruit other Elements to the Refractory ideologies, if only for a limited time. Elements like Copper and Iron have allied themselves with her to create high velocity penetrating shells as anti-tanks weapons.

W

Tungsten

74

J.Whyte

Rhenium 75

Name: Named after 'Rhenus' – the Latin name for the Rhine.

Subject Notes:

The information available on Subject 75 is limited as she is a rather misanthropic character. In fact despite painstaking research, information such as how she may affect soil, plants or animals is still ambiguous. Her feelings and attitudes toward humans are also a mystery. Though we do know she has no biological role, it is unknown as to whether she has a toxic outlook toward us. It is doubtful however, and if so her effects are thought to be minimal.

There are a few details that have been uncovered. They are as follows: she has the third highest melting point of all Elements (after Carbon and Tungsten at 3182 °C), the fourth highest density (after Osmium, Iridium and Platinum), and she has a high resistance to corrosion – all details that mark her clearly as one of the 'Refractory Metal Union'.

Madam Oxygen does have her effects on this recessive Element 'Oxidizing' her quickly, and Rhenium also shows evidence that she does not enjoy the moist English summer making her downcast and dull.

Subject History:

In 1925 Rhenium was finally discovered; the last stable, non-radioactive, naturally occurring Element to be found. As with all surreptitious undercover revolutionaries scientific authorities were anxious to find her, even interrogating Manganese until Element 75's whereabouts and secrets were unveiled. As there are no freely occurring signs of her and no minable mineral to be found, they resorted to looking into the movements of her comrades. As it happens she did and indeed does like to hide behind other Elements. She was eventually found lurking in the flue dust of Molybdenum and Gadolinite smelters.

After being discovered little was done with the newly found Element until the 1950's when she became more commercially viable in industry. Her work load includes contracts like working as a catalyst, or in coalition with other Elements for the common good in alloys, high temperature thermometers and very rarely in jewellery; though she disagrees with the general concept of useless adornment as a luxury of the bourgeois society.

Yet Rhenium seems to still enjoy her solitude, coming out for a hard day's work, and then quickly retreating to wherever she calls home.

75

RE

Rhenium

Sub-Chapter
The Platinum Metals

Platinum, slighted by the elite class of the Coinage Metals, and having disapprobation for the eminence given to certain Elements, decided the only course of action left to her was to create her own exclusive group, to raise herself to the prominence she feels she deserves; the Platinum group (abbreviated to the PGMs). With the heady tonic of spite, pride, rejection and a carefully nursed grudge she founded the PGMs (also known as 'The Platinoids' or 'Platidises'). These are the last of the Sub-Groups to be observed in this volume of 'The Illustrated Guide to the Elements'.

With a clear objective in mind, no small amount of megalomania, and wounded pride she went about selecting the Elements that would accompany her on her journey to an exclusory status. As for the strict guidelines the PGM clique must adhere to, it appears that all the Elements accepted into the group must have similar chemical and physical qualities to that of Platinum. Other necessities Element 78 deems important include being exceptionally rare, so as to add to their glamour and excellence and must be exceedingly dense (in fact the Elements in the PGM are the densest known Metal Elements). They must be highly tough and durable as well as fashionable, have stable electronic properties, be resistant to chemical attacks, have outstanding catalytic properties and (to give the PGM a decent, respectable name) they must be handsome, neat, à la mode and impervious to wear and tarnishing. Perhaps more important than all of these assets with respect to image, they must be above the common Elements and thus be of a Noble Metal standing. This amalgam of proficiency and charm ensure that the appearance of the group is stylish, sophisticated, yet has the skill to achieve their ultimate goal: a position of prominence.

Needless to say there were few Elements that could fit the high expectations of Platinum. In the end she did not have to look far to find the five other exceptional Elements to join her on her mission for prestige, for they all tend to dwell together in the same mineral deposits or ore bodies.

The six Elements we will be scrutinizing in this chapter are Platinum, Iridium, Osmium, Palladium, Rhodium and Ruthenium.

Ruthenium 44

Name: From the Latin word Ruthenia or Russia.

Subject Notes:

Ruthenium was the last of the Platinum Group Metals to be enlisted to the PGMs and one of the rarest Metals on Earth; as a result very little data has been collected about her.

The particulars that have been uncovered however are not encouraging. Element 44 is not a toxic Element in her own right, nor does she have any biological role, which at first glance seems reassuring but, on delving deeper, we expose a story far too common in the Elemental world. She is easily 'persuaded' to become toxic by the whiles of Oxygen ($RuO4$). The Oxide created by this abominable coalition is highly toxic to breathe. It seems plausible to assume Oxygen may hold some form of black-mailable material over this otherwise upstanding Element. After all is not unheard of for the well-to-do and recently affluent to have an ambiguous past.

Subject 44 holds all the hallmarks of a Platinum Group Metal: indifferent to most other Elements below her rank, hard, brittle and unaffected in the most part by air, water and acids. Ever aware of her image and how it may affect her delicate position in the PGM, she does not tarnish (unless extreme heat is applied - about 800°C), keeping lustrous shine and grace.

Subject History:

Because of her specialist skill set the demand for Ruthenium is ever on the increase. Her exceptional conductivity means that 50% of her effort goes into the electronics industry, and 40% into the chemical industry (for items like anodes for electronic cells giving them a longer life under corrosive condition). Added to Titanium she is able to make pipes corrosion resistant - perfect for undersea pipelines, and is renowned for doing her job with diligence, excelling at whatever she puts her hand to, once again enhancing the reputation of the PGM.

The above mentioned businesses may have earned her fortune but her favoured line of work is in jewellery production. She works closely with Platinum to make these objects of desire. She gives indications to suggest that she hopes, one day in the future, this will become her primary occupation.

44

RU

Ruthenium

Rhodium 45

Name: The Greek meaning 'rose coloured'.

Subject Notes:

We see a common occurrence in the Elements of the PGM, and that is a tendency toward vanity. Rhodium delights in her rose coloured crystals from which she gets her name, this pride making her no different from her 'Platinum' associates. She also possesses a skill when in a thin film, of becoming highly shiny and reflective like a mirror - a talent she uses to great effect in making sure, a hair is never out of place and her appearance immaculate.

However Subject 45 shows signs of insecurity regarding her place in the 'Platinum Group' hierarchy and as a result sometimes shows peculiar behaviour, either from strain or in an attempt to impress her peers. The most obvious example of this is when she becomes heated to her melting point (1963 °C) she begins to absorb Oxygen, but then to show how tenacious and independent she is she rejects Oxygen and then returns to her original solid state rather than becoming a Oxide minion. It may be that this insecurity stems from the knowledge she has the least 'bite' of the 'Platinum' circle; for she is the least toxic of all six Elements, so feels she must work all the harder to gain respect from Platinum herself. How much of her paranoia is justified is unknown.

Many details of this self-styled aristocrat remain elusive as she likes to control gossip that follows the well-tailored and rich. One method she employs is vigilance regarding where she frequents and with whom, she is notoriously difficult to fuse with other Metals, not wanting to sully her name with lesser Elements. Her presence will not be found in soil or in the atmosphere and she will only deign to appear in minuscular quantities in seawater. After all appearances and standards must be upheld.

Subject History:

Like most of the 'Platinum Group Metals' Rhodium comes from humble beginnings - namely as a by-product of Copper and Nickel mining, she has built herself into the PMG empire, albeit in a small way.

Her fortune was made by helping catalytic converters in cars, a job in which she excels as she is superb at reducing Nitric Oxide emissions. She also has her fingers in the glass industry, coating optical fibres, optical mirrors and reflective headlights.

RH ⁴⁵

Rhodium

Palladium 46

Name: Named after the asteroid 'Pallas'. This in turn was named after the Greek Goddess of Wisdom - Pallas.

Subject Notes:

Subject 46 is a rare Element and is often found in the company of the prosperous and prominent Elements, despite her chequered past concerning wealth and power. Regularly associating with Gold seems to have influenced the way she behaves as a Metal. Like Gold, she is capable of been beaten into extremely thin leaves, is unaffected by the beguiling of Oxygen, easy to work with even when cold (in fact this only makes her stronger and harder) and above all she has the attributes demanded of the higher classes of Metal: she is malleable and ductile.

Palladium is the quintessence of fine manners (despite her status of gentry and affinity close with the other, often haughty, PGM) and is a good natured Element and as far as we know has no history of toxicity or boorish behaviour.

Palladium has been named "the least noble" of the group because she is more willing to react with other Metals considered below her station than the other members of the Platinum group. She also holds the mantle for lowest melting point and least dense of the entire social group.

Yet despite her connections in high places, impeccable manners and well-to-do associates Palladium has struggled to gain any fame, power or genuine social standing.

Subject History:

The most intriguing aspect of Palladium is how she became known to us. She was first found with Platinum, and then in a mixture of Gold and Palladium where the combination was given the derogatory name of 'Useless Gold'.

William Wollaston first discovered and isolated this Element in 1803, but convinced there was more money to be made from this newly found Element did not announce his discovery. Rather he purified enough of the Element to sell it - without naming the discoverer - in a small shop in Soho. She was marketed as 'New Silver'; the Silver that would not tarnish in air and valued at six times more than the price of Gold. This however was only met with criticism; Richard Chenevix bought some of the 'new Silver' and dismissed it as an alloy of other Elements. In response to this snub Wollaston anonymously offered a reward of 20 British pounds (around £2000 today) to anybody who could produce 20 grains of the 'synthetic Palladium alloy' – needless to say nobody claimed the prize. It was not until February 1805 Wollaston revealed himself as the discoverer and, much to his disgust, he realised there was little profit to be made in this Element. The 'new Silver' ploy never took off and at the time the only uses that could be found for Palladium were in navigational instruments.

Today however Subject 46's luck has turned around, she mostly works in catalytic converters, electronics and in the chemical industry. But perhaps the most fitting job for a lady of her 'aspired to' status is in the jewellery trade. She has a beautiful shiny finish that does not tarnish easily. It is these properties make her so desirable in jewellery and objects of art. She is now marketed as 'White Gold'.

46

PD

Palladium

Osmium 76

Name: From the Greek 'osme' meaning smell, because the Metal surface gives off a volatile Osmium Trioxide which has an irritating and pungent odour.

Subject Notes:

Throughout our investigation we have seen Elements of a despicable, degenerate and unpleasant nature, but there a there are few who are quite as obnoxious, contemptible and unreasonable as Osmium. She bears traits in common with the abhorrent Arsenic (volume i). If one recalls Arsenic on her own is not virulent, only when paired with Oxygen does her poisonous nature reveal itself. The same is true of Element 76, the mephitic Osmium tetroxide ($OsO4O$) is highly volatile, penetrates skin readily, and is intensely toxic by inhalation, ingestion, and skin contact. Airborne concentrations as low as 1% of Osmium tetroxide vapour can cause lung congestion and skin or eye damage, but unlike Arsenic or other poison Elements she has no villainous objectives or inherent chemical conditions. It is purely bad manners.

This is not the end of her vexing character flaws; Osmium is immensely difficult to work with even without her foul odour and noxious moods. As the densest Metal known to man and has one of the highest melting points of the Elements, she is difficult to shape and it is impossible to use Subject 76 for anything she does not wish to do. High temperatures and pressures are required to force her hand.

She has few friends and therefore keeps Iridium close by her, indeed it is hard to separate the two. As Iridium is incredibly fashion conscious and PGM code demands it, Osmium is forced to make an effort and will maintain her lustrous glow even when heated to extremely high temperatures. Her unfortunate choice of perfume is, therefore, a shame.

Osmium's saving grace is that she is a tough metal and thankfully a rare one.

Subject History:

At this juncture one may be asking how she came by her wealth and status if she is unwilling to earn her place? The answer lies in her work which comprises of small jobs in the chemical industry, in compass needles, in gramophone needles, clock bearings and most famously in high quality fountain pens (partnered with Iridium); jobs where extreme hardness and resistance to corrosion are needed.

There is another job of interest which we shall look at deeper, that of forensic chemistry. Osmium tetroxide has been used in fingerprint detection. As a strong Oxidant it cross-links lipids, or fats, left by the fingers by reacting with unsaturated Carbon-Carbon bonds, and thereby fixes both biological membranes in place in tissue and simultaneously stains them. Because Osmium atoms are extremely electron dense, Osmium staining greatly enhances image contrast in transmission electron microscopy (TEM) studies of biological materials.

S

76

osmium

Iridium 77

Name: Named after Iris the Greek Goddess of the rainbow, because her salts are of such varied colours.

Subject Notes:

Iridium enjoys the high life her association with the Platinum Group provides and largely spends her time with those inside the assemblage, most notably her closest companion: Osmium. Element 77 has shown a flair for fashion and would rather be seen dead than in an outdated gown or démodé dress. As a direct result Subject 77 is in the possession of hundreds of outfits in every colour and style for any occasion, hence we see how she obtained the name Iridium and why many in the Element world look to her for the latest styles.

But what of her character outside of the realm of fashion? Close observations show a hard hearted, brittle-tempered Element. Indeed the most resistant of all Elements known to us, beating Osmium by the finest of margins. Her closeness to Osmium (they are hardly ever apart), means she has picked up many of her characteristics. She however is nowhere near as insufferable, particularly in the nasal area and in Osmium's toxic attitude toward most things, though she still has an unpleasant parvenue prejudice. She prefers the company of those on the same social level (i.e the Platinum Group) and tries to avoid the plebeian Elements, remaining a stylish yet inert Element.

Subject History:

As with all of the Platinum Group Metals subject 77 is rare and gathered as a result of Platinum refining. As a refined Element she has a few roles, but roles she was born to fill. Her professions allow her incorruptibility and outstanding toughness to be used to her maximum ability in compass bearings, long life engine parts, deep water pipes, and her project with Osmium regarding fountain pen nibs.

However one of the most intriguing findings about this Element is her astronomical beginnings. Worldwide there is a layer of Iridium. This layer of residue has been there since the end of the Cretaceous period. Since meteors and asteroids contain a higher percentage of Iridium than the earth's crust, this Iridium enriched deposit is seen as evidence by many that the earth was struck by a hefty sized meteor or asteroid at around about that time. Dust from the impact would have spread around the globe, depositing the Element in question. The theory continues that as a result the dust would have blocked the sun for an extensive time, resulting in the extinction of dinosaurs. This has also helped our planet to be enriched by the Element, though she is still one of the rarest of Elements in the earth's crust.

Platinum 78
Name: From the Spanish 'platina' meaning 'Little Silver'.

Subject Notes:

Some Elements are born into wealth and fame, some are lifted to prominence by their deeds, Subject 78's notoriety was by neither of these courses but by pure perseverance, cunning and no small amount of stubbornness.

Today, Platinum enjoys the prestige of a 'Metal of luxury', and thanks to good marketing is seen as more popular and expensive than Gold. Surrounding herself with glamour she has become invaluable in the fight against cancer. Cancerous cells will not divide on her, she can thus put a stop to the spread, though not without cost.

Element 78's record has not always been one of adulation. When the Spaniards conquered Ecuador and Colombia in the 16th century they discovered this Metal which was like Silver, but twice as dense, and had a pearlescent shine, but they were not interested as they only had eyes for Gold. Undeterred by this minor setback Subject 78 waited. It was not until the 18th Century that interest was shown in her and her qualities. Like an incorruptible Silver that never tarnishes and is always immaculately dressed, this Element is unaffected by air, water, or corrosion and is so self-confident and powerful that Oxygen has no hold on her. Platinum started to be shipped to Spain; the first leg of her mission was complete, making it to Europe.

Once making it to Europe she was met with opposition as the Spanish banned her, under the charge of the adulteration of Gold. Her mines were closed and masses of her stocks were disposed of into the sea. Still Subject 78 was not disconsolate, outrage kept her fighting for what she believed was her right. Storing up all the negativity for a date when she was better suited to exact her revenge, she bided her time.

Platinum came back into favour, yet was thought to be as common as Silver and Gold and in Russia coins began to be issued made of her. This was the start of the rise for Element 78, but was not enough to satisfy her. After the production of around 1.5 million coins the project was abandoned, as the price of the Metal began to outweigh the value of the coin.

Platinum is a difficult Metal and not of a favourable disposition. Years of disappointment have made her hard and indifferent, her high Melting point, density, and toughness made her hard to work with, which only made the people more infatuated. By the 19th Century we had finally worked out a compromise with the haughty Metal, making items of beauty and practicality. At last her empire was taking shape.

The price for the luxury or health provided by Subject 78 and her cronies, can be high. The majority of people who come into contact with this highly intelligent yet rancorous Element suffer from 'platinosis' an allergic reaction with symptoms akin to the common cold. Instances are rare as she is an "exclusive Woman of means" and few are exposed to compounds of Platinum which would cause the reaction. Those who suffer from cancer also suffer the effects of the Platinum in their system.

In evaluation of this Element we see that though her personality has some undesirable aspects, she is a genius in her own way - persevering and a true visionary. Credit must be given to her ambition and ability to see it through.

78
PT
Platinum

Notes

The Periodic Table of Elements

1	2	3	4	5	6	7	8	9	10	11	12	13	14	15	16	17	18
H 1 Hydrogen																	HE 2 Helium
LI 3 Lithium	BE 4 Beryllium											B 5 Boron	C 6 Carbon	N 7 Nitrogen	O 8 Oxygen	F 9 Fluorine	NE 10 Neon
NA 11 Sodium	MG 12 Magnesium											AL 13 Aluminium	SI 14 Silicon	P 15 Phosphorus	S 16 Sulphur	CL 17 Chlorine	AR 18 Argon
K 19 Potassium	CA 20 Calcium	SC 21 Scandium	TI 22 Titanium	V 23 Vanadium	CR 24 Chromium	MN 25 Manganese	FE 26 Iron	CO 27 Cobalt	NI 28 Nickel	CU 29 Copper	ZN 30 Zinc	GA 31 Gallium	GE 32 Germanium	AS 33 Arsenic	SE 34 Selenium	BR 35 Bromine	KR 36 Krypton
RB 37 Rubidium	SR 38 Strontium	Y 39 Yttrium	ZR 40 Zirconium	NB 41 Niobium	MO 42 Molybdenum		RU 44 Ruthenium	RH 45 Rhodium	PD 46 Palladium	AG 47 Silver	CD 48 Cadmium	IN 49 Indium	SN 50 Tin	SB 51 Antimony	TE 52 Tellurium	I 53 Iodine	X 54 Xenon
CS 55 Caesium	BA 56 Barium	LU 71 Lutetium	HF 72 Hafnium	TA 73 Tantalum	W 74 Tungsten	RE 75 Rhenium	OS 76 Osmium	IR 77 Iridium	PT 78 Platinum	AU 79 Gold	HG 80 Mercury	TL 81 Thallium	PB 82 Lead	BI 83 Bismuth	PO 84 Polonium	AT 85 Astatine	RN 86 Radon
FR 87 Francium	RA 88 Radium																

Lanthanoids

LA 57 Lanthanum	CE 58 Cerium	PR 59 Praseodymium	ND 60 Neodymium	PM 61 Promethium	SM 62 Samarium	EU 63 Europium	GD 64 Gadolinium	TB 65 Terbium	DY 66 Dysprosium	HO 67 Holmium	ER 68 Erbium	TM 69 Thulium	YB 70 Ytterbium

Actinoids

AC 89 Actinium	TH 90 Thorium	PA 91 Protactinium	U 92 Uranium	PU Plutonium

The Basics of Elemental Physiology

To understand the Elements one first must understand their biology. Humans have DNA which gives them characteristics that they pass down as family traits. Elements have individual atomic structures.

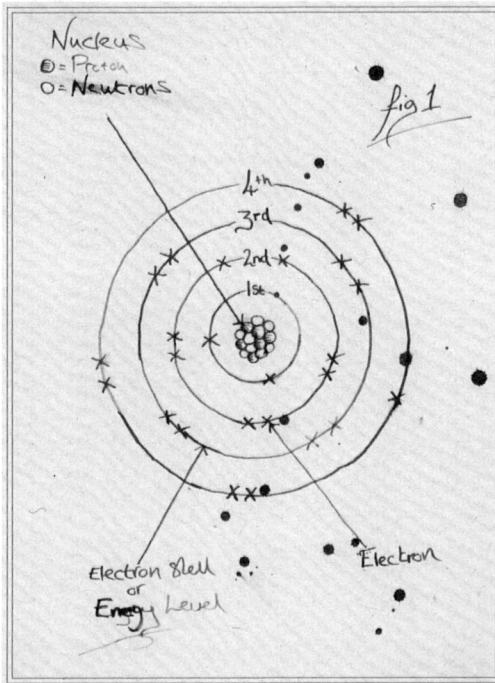

Nucleus
● = Proton
○ = Neutrons

fig 1

4th
3rd
2nd
1st

Electron Shell
or
Energy Level

Electron

The atomic structure of an Element dictates her size, the state which she lives and how 'electron hungry' she is.

An atom is made of three main parts: Protons, Neutrons and the ever famous Electrons. Protons and Neutrons form the Nucleus (See fig. 1).

Protons are positively charged, and thus attract the negatively charged Electrons. Neutrons are neutral.

Electrons move around the Nucleus in shells and have very little mass. Electrons and Shells are the very base of Chemistry as we know it. The number of Electrons in an Atom is always equal to that of Protons. However, an Element is only happy (i.e. stable) when all of its shells are full. In most atoms this is rare; and this is the cause of reactions.

Electrons can be freely traded between Atoms, and Elements have different ways of getting what they want, gaining, losing or sharing electrons. When gaining or losing an electron they become charged one way or the other.

How reactive an Element is, is dependent on how far away the electrons are from the nucleus, and how full or empty their shells already are.

Glossary of Terms

Creep: The tendency of a solid material to slowly move or deform permanently under stress. All ways increases with higher temperatures.

Electronegative: The more electronegative the more powerfully it attracts electrons.

Refractory Metals: Metals with a very high resistance to heat and water.

Tensile Strength: The force required to pull to breaking point.

'Aqua Regia': "King of Waters": mixture of Nitric and Hydrochloric Acids.

CNO Cycle: Or Carbon- Nitrogen- Oxygen cycle. It is a catalytic cycle self sustaining, and the dominant source of energy in the stars.

Caesium Clock: A type of atomic clock that uses the frequency of radiation absorbed in changing the spin of electrons in caesium atoms.

Blue Stocking Woman: An intellectual or literary woman. Originally from 'The Blue Stockings Society' which was an informal women's social and educational movement in England in the mid-18th century.

Borate/ Boride: A salt or boric acid/ a compound in which Boron is the most electronegative element.

Pyrophoric: A pyrophoric substance is a substance that will ignite spontaneously in air.

Phylum: Family or Group.

Helium II: Liquid Helium existing as a superfluid below the transition point of approximately $2.2°K$ at 1 atmosphere and having extremely low viscosity and extremely high thermal conductivity.

Ferromagnetism: is the basic mechanism by which certain Elements form permanent magnets, or are attracted to magnets.

Alloy: An alloy is a mixture composed of two or more elements.

Index